Verónica del Pilar Gavilanes Fray
Fausto Mauricio Tamayo Vásquez
Blanca Georgina Costales Coronel

Empty Bodies, Limited Minds

Verónica del Pilar Gavilanes Fray
Fausto Mauricio Tamayo Vásquez
Blanca Georgina Costales Coronel

Empty Bodies, Limited Minds

Malnutrition and its Impact on the Psychomotor Development of Children

ScienciaScripts

Imprint

Cover image: www.ingimage.com

This book is a translation from the original published under ISBN 978-613-9-44082-5.

Publisher:
Sciencia Scripts
is a trademark of
Dodo Books Indian Ocean Ltd. and OmniScriptum S.R.L publishing group

120 High Road, East Finchley, London, N2 9ED, United Kingdom
Str. Armeneasca 28/1, office 1, Chisinau MD-2012, Republic of Moldova, Europe
Printed at: see last page
ISBN: 978-620-8-26134-4

"Empty Bodies, Limited Minds: Malnutrition and its Impact on Child Psychomotor Development".

Authors: Verónica Gavilanes Fray, Mauricio Tamayo Vásquez, Blanca Costales Coronel.

SUMMARY

Malnutrition in children under five years of age has various causes and consequences that have an impact at the national level. This research aims to contribute to the reduction of child malnutrition. Childhood is a crucial stage in human development, marked by growth and development that requires adequate nutrition, essential for the body. In Ecuador, 27.2 per cent of children under the age of two suffer from chronic malnutrition, which affects the country's productivity and has repercussions throughout life. The lack of healthy nutrition before the age of 5 negatively influences the physical, emotional and intellectual development of children. Undernourished children often have lower weight and height than expected for their age. However, there are multiple factors that contribute to child malnutrition, including brain damage caused by lack of nutrients. Moreover, the problem is not limited to food deficiency, but may be related to parents' lack of knowledge about proper nutrition for their children, which can have adverse effects on the body.

Introduction to Child Undernutrition

- **Definition of malnutrition**

Nutrition is an essentially biological process that is activated when humans consume food necessary to maintain a healthy life. This process involves the absorption of essential nutrients, which are integrated into the body to ensure proper growth and optimal functioning of all bodily functions. The quality of nutrition not only impacts physical development, but also overall well-being, including mental and emotional aspects. Balanced nutrition is essential for strengthening the immune system, improving cognitive ability and preventing long-term illness. In addition, adequate nutrition during critical stages of growth, such as childhood and adolescence, is crucial for the overall development of the individual, influencing their future health and quality of life. (Carpio-Arias, et al., 2018)..

According to (Martínez, Salazar Duque, Portugal Morejón, & Lala Gualotuña, 2020)malnutrition refers to a set of clinical symptoms, as well as biochemical and anthropometric alterations that result from insufficient intake and/or inadequate use of macronutrients, which leads to the unsatisfaction of the body's nutritional needs. This condition not only reduces the body's defences, increasing vulnerability to various diseases, but also manifests itself in a body weight or height below what would be appropriate for the individual's age.
Undernutrition is therefore a complex and multifactorial phenomenon, influenced by a wide range of social and economic conditions. It is one of the most serious problems facing children globally. According to UNICEF, the World Health Organization (WHO) and the World Bank (2019), child undernutrition is a reflection of structural inequalities that affect millions of children

around the world, limiting not only their physical development, but also their cognitive potential and their ability to lead healthy lives in the future. This problem demands urgent attention and a comprehensive approach that addresses both the immediate and underlying causes of malnutrition, including poverty, lack of access to health and education services, and food insecurity. (UNICEF, 2022), (Health, 2024).

Definition of malnutrition

Malnutrition can be defined as a nutritional imbalance that manifests itself in an individual, either through a deficient or excessive diet. This condition is one of the most important factors contributing to the global burden of disease, as more than one-third of childhood illnesses worldwide originate from malnutrition, which is mainly triggered by poverty, according to (Gonzalez, Font, Ochoa, Rodriguez, & Estrada, 2020).

Malnutrition has historically been recognised as an essential component of maintaining good health and ensuring an adequate quality of life. Since ancient times, figures such as Hippocrates and Galen considered it to be of vital importance. Protein-energy malnutrition, in particular, refers to a set of clinical manifestations, biochemical and anthropometric alterations that occur when a person does not ingest the necessary amount of nutrients according to their weight, which their organs and tissues need to function and develop properly. This condition is known as a systemic alteration that is potentially reversible and presents different degrees of intensity. (Cuevas-Nasu, et al., 2019)..

It is now recognised that, since the second half of the last century, new research and significant changes have emerged in the field of nutrition, posing important theoretical challenges. Malnutrition is essentially a multifactorial phenomenon, involving biological, social, cultural and socio-economic aspects. (Paraje, 2008).. Furthermore, (Pozo Aguilar, 2022) points out that this pathology, being the result of insufficient food consumption, facilitates the sudden onset of infectious diseases. Malnutrition is classified according to its time of onset as acute or chronic; according to its cause, it can be primary, if it is due to a lack of food intake, or secondary, if it is due to a biological impact caused by other pathologies. It is also classified according to its severity as mild, moderate or severe, and according to its pathophysiological mechanism, as marasmus, Kwashiorkor or mixed.

This holistic and complex approach to malnutrition underscores the need to address it from multiple perspectives, considering not only the provision of food, but also the socio-economic and cultural conditions that contribute to its development. The challenge of addressing malnutrition globally calls for integrated solutions that can be adapted to different contexts and needs, recognising that it is a problem that profoundly affects the physical, mental and social well-being of people, especially the most vulnerable.

Types of malnutrition

According to (Cancela, 2021)malnutrition manifests itself as a disease resulting from an inadequate diet, low in calories and protein. This condition can be caused by social factors, such as famines affecting underdeveloped regions, or by deficiencies in the body, as in the case of anorexia, where the body is unable to absorb the necessary nutrients. Malnutrition has a particularly devastating impact during critical stages such as pregnancy, foetal phase, infancy, childhood and old age. It currently affects one in six adults and one in three children.

This situation reflects not only the lack of access to adequate food, but also the complexity of the factors contributing to undernutrition, which include both biological and socio-economic elements. During pregnancy and the foetal phase, undernutrition can cause severe problems in foetal development, increasing the risk of complications for both mother and child. In childhood, the condition can result in poor growth and cognitive problems, which often have lasting effects on health and well-being throughout life.

Similarly, undernutrition in old age can lead to a more rapid deterioration of health, aggravating chronic diseases and increasing vulnerability to infections. The prevalence of undernutrition at these stages of life underlines the importance of addressing it not only as a food problem, but as a challenge that requires comprehensive and sustainable interventions. Promoting policies and programmes that ensure access to adequate nutrition for all people, especially the most vulnerable populations, is essential to mitigate the devastating effects of undernutrition at all stages of life.

Table 1. Types of malnutrition

TYPE OF CHRONIC CHILD MALNUTRITION	FACTORS INFLUENCING CHILD UNDERNUTRITION	CHILD MALNUTRITION DEMONSTRATIONS
CHRONIC CHILD MALNUTRITION	-Stunting. -It indicates a deficiency of necessary nutrients over a prolonged period of time.	It is said to be a multifactorial phenomenon, resulting from a wide range of social and economic conditions.
MODERATE ACUTE MALNUTRITION IN CHILDREN	-It weighs less than is appropriate for its height. -It is also measured by arm circumference, which is below the reference standard. -It requires immediate treatment to prevent it from getting worse.	It is manifested by low weight in relation to the height of the individual, which is caused by a recent situation of lack of food or an illness that has produced a rapid weight loss. This type of malnutrition is recoverable, but if it is not treated in a timely manner, it puts the individual's life at high risk.
SEVERE OR SEVERE ACUTE MALNUTRITION IN CHILDREN	-The child is well below the reference standard for his or her height. -It is also measured by arm circumference. -It disrupts all of the child's life processes and carries a high risk of mortality. -The risk of death for a child with severe acute malnutrition is 9 times higher than for a normal child.	It is manifested by a low height according to the age of the individual, as a consequence of recurrent diseases and/or a deficient and prolonged dietary intake. This type of malnutrition permanently diminishes the physical, mental and productive capacities of the individual, when it occurs between gestation and thirty-six months.

The Marasmus

This type of malnutrition is characterised as the result of low food intake and is especially common among children from lower social classes in urban areas. These children are often breastfed for a short period of time and then fed diluted and often water-contaminated milks, which increases the risk of diarrhoea and gastrointestinal infections, contributing to their malnutrition.

An infant suffering from marasmus, a severe form of malnutrition, shows a significant delay in weight for age and weight for height. In addition, they lose much of their subcutaneous fatty tissue, leaving their skin flabby and their bones more visible, giving them a frail and emaciated appearance. Not only do these children experience stunted physical growth, but they may also suffer from muscle weakness, impaired cognitive development and a compromised immune system, making them more vulnerable to other diseases.

The situation is particularly acute in contexts where poverty limits access to adequate nutrition and health services, perpetuating a cycle of malnutrition and disease. For these children, early intervention is crucial, as prolonged undernutrition can have irreversible effects on their physical and mental development. Addressing this form of malnutrition requires not only improving access to safe and nutritious food, but also implementing education and family support programmes to ensure that children receive adequate care and nutrition from birth. (Saca Molina, 2018).

(Realpe Muñoz, 2013) describes marasmus as a chronic form of protein-energy malnutrition (PEM), characterised by a severe loss of subcutaneous tissue, muscle and fat, causing affected children to have an aged appearance or wrinkled skin. Despite the severe wasting these children suffer, they are surprisingly better able to withstand reduced protein and lipid intake compared to other types of malnutrition.

However, this resilience should not be interpreted as a sign of health, but as the body's adaptation to extreme nutrient deficiency. Although marasmus may appear less deadly in the short term than other forms of malnutrition, such as Kwashiorkor, its long-term effects are equally devastating. Children with marasmus not only experience significant physical growth retardation, but also face an elevated risk of infections and cognitive impairment as their body, in an effort to conserve energy, sacrifices the development of vital functions.

Early intervention is essential for these children, as marasmus, if not properly treated, can lead to serious complications and even death. It is crucial to provide intensive nutritional support and medical care to reverse the effects of this condition and allow children to regain their normal health and development. In addition, it is important to address the underlying causes of marasmus, such as poverty, food insecurity and lack of access to health services, to prevent more children from suffering from this severe form of malnutrition.

Kwashiorkor

This type of malnutrition is characterised by an almost exclusively protein deficiency. It usually affects children who, after a prolonged period of breastfeeding, start receiving a diet based mainly on carbohydrates but with a very low protein intake. These children have oedema in the abdominal area, and their state of malnutrition may be worsened by various infections. (Naranjo Castillo, Alcivar Cruz, Rodriguez Villamar, & Betancourt Bohórquez, 2020)..

Clinically, one of the most distinctive signs is the presence of oedema, accompanied by other symptoms such as dermatosis, diarrhoea, fatty liver, hair changes, loss of interest and decreased appetite. (Realpe Muñoz, 2013). Despite malnutrition, the child's weight may appear to be adequate for his or her age. This type of protein-energy malnutrition (PEM) is more common in rural areas of developing countries, mainly affecting children around 2 years of age, coinciding with the weaning period and the introduction of a diet rich in starch but poor in protein.

This type of malnutrition is associated with protein loss in the visceral compartment of the body, especially the liver, while muscle and adipose tissue are usually relatively preserved. The resulting hypoalbuminaemia leads to generalised oedema, which may mask actual weight loss. The enlarged fatty liver is caused by inadequate lipoprotein synthesis, leading to accumulation of triglycerides in the liver, mobilised from other parts of the body. In addition, affected children often show apathy, listlessness and anorexia. Reversible atrophy of the mucosa of the small intestine can lead to malabsorption problems. (Connect, 2018).

This type of malnutrition highlights the importance of a balanced diet that not only provides energy but also essential nutrients, such as protein, for a child's overall health and development. Without proper intervention, this condition can have devastating long-term effects on the health and development of affected children.

Degrees of child malnutrition

Table 2. Degrees of child undernutrition

Degrees of child malnutrition		
Undernutrition Grade I	**Malnutrition Grade II**	**Malnutrition Grade IIII**
Grade I or mild undernutrition is when the weight for age of the child is normal, but the weight for height is low. These are children who, despite having a normal height, have not been able to reach a weight for height.	Grade II or moderate undernutrition occurs when a child under one year of age has a low weight-for-age. Moderate undernutrition is also considered when children aged 1-4 years have a low weight-for-height ratio.	Grade III or severe undernutrition occurs if the child under one year of age has a deficit of 40 per cent or more of the ideal weight for his or her age. In addition, severe undernutrition occurs when a child over one year of age has a reduction in the weight/height ratio of more than 30 per cent.

According to (UNICEF, 2021) it is important to clarify that the earlier malnutrition is treated, the faster the recovery and the lesser the after-effects that may remain.

Types of determinants of undernutrition

There are different types of determinants of undernutrition, which can be classified as immediate, underlying and basic. Immediate determinants include factors such as lack of breastfeeding until an appropriate age and the presence of infectious diseases, such as acute respiratory infections and acute diarrhoea, which affect both nutrient intake and utilisation. In addition, undernutrition is also influenced by insufficient diets, both in quantity and quality, reflecting a lack of adequate complementary feeding (Fernández-Martínez, 2022). (Fernandez-Martinez, 2022).

Underlying determinants of undernutrition include food insecurity, which refers to lack of consistent access to nutritious and sufficient food; inadequate child care and feeding practices; lack of access to medical services; and poor hygiene conditions, which can exacerbate vulnerability to disease and thus to undernutrition. On the other hand, underlying determinants are linked to broader factors, such as the country's economic and social policies, the socio-cultural environment and the resources available at the community and national level (Fernandez-Martinez, 2022)..

The complexity of undernutrition lies in the interaction between these different levels of determinants. Immediate factors, such as disease and poor nutrition, are often symptoms of deeper underlying and basic problems. For example, food insecurity is not only the result of poverty, but also of inadequate policies that do not ensure equitable distribution of resources. Similarly, lack of access to health and education services can be the result of political and economic decisions that do not prioritise the well-being of the most vulnerable.

To effectively address undernutrition, it is crucial to implement a comprehensive approach that not only treats the immediate symptoms, but also addresses the underlying and basic causes. This includes improving food security, promoting appropriate care and feeding practices, ensuring access to quality health services and promoting policies that reduce social and economic inequalities. Only through a multifaceted and sustained approach can the prevalence of undernutrition be reduced and the health and well-being of affected populations be improved.

Global and regional statistics

Childhood undernutrition is recognised as a global and widespread problem that arises due to a deficiency of essential nutrients in children's diets. This condition not only affects an infant's immediate well-being, but also has long-term consequences, including significant delay in physical and cognitive development, and in the most severe cases, the onset of severe and debilitating diseases.

The impact of undernutrition in the early years of life is of particular concern, as this stage is crucial for growth and brain development. Children who do not receive adequate nutrition can experience slow growth, impaired immune system development and increased susceptibility to infections and chronic diseases. In the long term, these effects can perpetuate a cycle of poverty and ill health, affecting not only individuals but also the socio-economic development of entire communities.

In addition, childhood malnutrition is closely linked to poor academic performance, which can limit educational and employment opportunities in adult life, thus perpetuating inequality and social exclusion. In the most extreme cases, malnutrition can lead to catastrophic medical conditions such as Kwashiorkor or marasmus, which, if not properly treated, can be fatal.

Addressing child undernutrition therefore requires a multifaceted approach that includes improving access to nutritious food, promoting breastfeeding, educating about healthy eating practices, and implementing public policies that address underlying socio-economic inequalities. Only through a concerted and sustained effort can the prevalence of child undernutrition be significantly reduced and its devastating long-term consequences mitigated. (Ramos-Padilla, 2020)..

According to the World Health Organization (WHO), in 2018, approximately one-third of all child deaths were attributed to undernutrition. This condition remains one of the leading causes of ill health and premature mortality in children under five years of age, especially in developing countries. (Health O. M., 2021)..

Malnutrition is a complex and multifactorial problem that not only stunts children's physical growth, but also negatively affects their cognitive and emotional development. Various social, cultural and economic factors contribute to the high prevalence of child undernutrition, and these causes are most prevalent in rural communities. In these areas, poverty, lack of access to nutritious food, inadequate feeding practices and lack of basic health services create an environment in which children are particularly vulnerable to undernutrition.

The impact of undernutrition on child development has long-term consequences, affecting not only individual health, but also the economic and social progress of affected communities. Addressing this problem therefore requires comprehensive interventions that consider not only the provision of food, but also education, improved living conditions, and the implementation of policies that address the structural inequalities that perpetuate undernutrition in the most vulnerable regions. (UNICEF, 2021).

Chronic child malnutrition affects 27.2 per cent of children under the age of two in Ecuador, making the country the second most malnourished in Latin America, behind only Guatemala. This situation has profound repercussions on national productivity and a lasting impact on people's lives. According to (Quirindumbay Uchupailla, 2016) the lack of adequate nutrition in children under 5 drastically affects their physical, emotional and intellectual growth. Children suffering from malnutrition generally do not reach the weight and height corresponding to their age, which translates into insufficient physical growth and very low body weight.

Undernutrition is a serious global problem due to its wide magnitude and its impact on child morbidity and mortality, especially in contexts with unfavourable socio-economic conditions that deteriorate the quality of life of the general population. (Cuevas-Nasu, et al., 2019)..

This nutritional deficiency can originate from the prenatal period, affecting children from gestation through the first three years of life, leading to primary and chronic undernutrition. These conditions have a significant influence on the physical, intellectual, emotional and social capacity of children, as nutritional deterioration reduces their ability to learn during the school years and limits their opportunities to access higher levels of education. (Jiménez Benítez D., 2010)..

However, child undernutrition should not be seen solely as a problem of lack of food, but as a deeper social issue that requires holistic consideration when seeking solutions for those most affected. (Rivadeneira, 2020)..

In addition, a distinction can be made between primary and secondary undernutrition; primary undernutrition arises when there is a lack of nutrients in the child due to various established factors, while secondary undernutrition is associated with pre-existing diseases in the child, which are not dependent on social or cultural factors. These diseases may include genetic, immunological, and other pathologies that cause deformities in organs such as the heart, brain, liver and kidneys. Secondary malnutrition can be identified through a medical assessment, as it presents signs and symptoms such as diarrhoea, weight loss, dehydration and infections. (Cueva Moncayo & Pérez Padilla, 2021)..

It is essential to promote and support breastfeeding as a key strategy to ensure proper child development and prevent both child undernutrition and various infectious diseases. Exclusive breastfeeding (EBF), which involves feeding the baby only breast milk, excluding solids and other liquids, has been shown to be highly effective in reducing the risk of disease.

Studies have shown that breastfeeding reduces the incidence of infections, particularly gastrointestinal infections, with a 64% reduction in their occurrence. Moreover, the protective effect of breast milk is not only observed during breastfeeding, but persists for up to two months after breastfeeding has stopped, providing an additional barrier against infections and other health problems. Therefore, the promotion of breastfeeding is essential not only for the physical well-being of the child, but also for improving the child's immune defences over time. (Brahm & Valdés, 2017)..

Causes of Child Undernutrition

The contextual factors that influence health inequalities are manifold and cover different levels of influence, ranging from governmental decisions to social and cultural aspects. Among the most important of these are governments, macroeconomic policies, and health and social policies, which play a key role in the distribution of health resources. In addition, factors such as culture, values and social norms also contribute to shaping inequalities in access to health. In particular, the welfare state and redistributive policies emerge as two of the most influential elements in ensuring better public health; the World Health Organisation (WHO) also highlights the relevance of social hierarchy, which defines class relations and social structure in communities. (Health O. M., 2024). .

According to (Alvarez Ortega, 2019) one of the key determinants of health inequalities is socioeconomic position, which is linked to factors such as income, access to resources, and employment. These structural determinants, such as gender, ethnicity or race, directly affect health, creating a social stratification that generates disparities in living and working conditions. This stratification reinforces social inequalities through factors such as education, occupation and other structural determinants.

On the other hand, (Brahm & Valdés, 2017) points out that direct causes of child stunting include inadequate breastfeeding practices, insufficient complementary feeding and the incidence of infectious diseases. Poor energy and nutrient intake results from inappropriate feeding practices and low-quality diets, especially in extremely poor populations, where household food insecurity is common.

Child malnutrition is caused by poor nutrition in quantity and quality in both mothers and children, as well as lack of access to safe drinking water and unsanitary conditions, which lead to infectious diseases such as diarrhoea. To combat this problem, it is crucial to implement effective food policies focused on the most vulnerable sectors. Failure to do so will result in stunted physical and intellectual growth, leading to adults with limited productivity and a propensity to develop chronic, cardiovascular and metabolic diseases.

According to the author (Santafé Sánchez, Sánchez Rodríguez, A.L., & C.H., 2012) poor food intake, in which the human body expends more energy than the food it consumes. There are medical illnesses that have the potential to trigger malabsorption or problems in food intake thus causing malnutrition. Oh social, environmental or economic situations have the potential to drag individuals into malnutrition. These reasons have the potential to be:

Medical pathologies: Anorexia nervosa, Bulimia, Celiac disease, Coma, Depression, Diabetes mellitus, Gastrointestinal disease, Constant vomiting.

Complications

Heart: the heart loses muscle mass, as well as other body muscles. In the most advanced stage there is heart failure and death follows.

Immune system: becomes inefficient. The body cannot make guard cells. Afterwards, infections of the intestine, respiratory, and other events are common. The duration of pathologies is longer and the prognosis consistently worse than in typical individuals. Healing is slowed.

Blood: iron deficiency anaemia related to malnutrition is likely to occur.

Intestinal tract: there is less secretion of HCL by the stomach, making that environment more suitable for bacterial growth. The intestine reduces its peristaltic rate and its absorption of nutrients is quite limited. (Senna Rodrigues, Campos Pellanda, & Andreatta Gottschall, 2012)..

Socio-economic factors

According to (Alvarez Ortega, 2019)low family income almost always accompanies malnutrition, which leads to low availability and access to food, lack of means to produce or buy it, poor sanitary conditions, poor infant care, lack of access to education, poor feeding practices, dietary whims and emotional factors.

- Socio-cultural factors

The role of the family is crucial in the care and support of child nutrition, as stated by (Alvarez Ortega, Desnutrición infantil, una mirada desde diversos factores, 2019).. Studies show that the father, mother and other close family members, such as the mother-in-law, play a significant role in the healthy development of the child. This support is not only limited to providing food, but also to ensuring an adequate caring environment, including the time spent with the child and the socio-cultural conditions in which the child grows up.

Family involvement in decisions about children's nutrition and well-being contributes significantly to the prevention of child undernutrition. A family environment that promotes healthy eating habits, as well as a nurturing and protective environment, can counteract external factors that negatively influence children's health, such as poverty or lack of access to basic health services. For this reason, it is important to highlight that comprehensive family support, especially in the first years of life, is a determining factor in guaranteeing the adequate growth and development of the child, considerably reducing the risks of malnutrition and other nutritional deficiencies.

- **Biological factors**

Biological factors refer to those elements inherent in an individual's susceptibility to develop undernutrition. These factors influence the body's ability to adequately utilise nutrients, beyond the quantity or quality of food consumed. In other words, even if a person eats sufficient or good quality food, the presence of biological deficiencies can interfere with the proper absorption and utilization of nutrients, limiting nutrient utilization and increasing the risk of malnutrition. These limitations may be related to aspects such as metabolism, enzyme problems, or deficiencies in key vitamins and minerals, highlighting the importance of considering both the quantity and efficiency with which the body processes food. (Álvarez Ortega, Desnutrición infantil, una mirada desde diversos factores, 2019)..

- **Environmental factors**

  According to (Martínez & Andrés, 2007)environmental factors determine the environment in which children and their families develop. These factors include both natural risks inherent to the environment, such as natural disasters (floods, earthquakes, hurricanes, etc.), and those generated by human activity, also known as anthropogenic factors, which include problems such as water, air and food pollution. These environmental factors have a direct bearing on child undernutrition, as the quality of food production and the occurrence of epidemics are profoundly influenced by the state of the environment. The vulnerability of children under five is particularly critical, as their developing immune systems are not strong enough to resist diseases that thrive in adverse environmental conditions.

In addition, the impact of environmental factors not only affects the availability of healthy food, but also increases the risk of infectious and parasitic diseases, which further aggravate the situation of undernutrition. Rural and impoverished areas, where basic services are limited or non-existent, tend to be the most affected, exacerbating inequalities in child health and development.

- **Housing environmental factors**

  (Chimborazo Bermeo & Aguaiza Pichazaca, 2023) points out that environmental housing conditions also play a crucial role in children's well-being, representing a significant barrier to children's healthy development. Lack of access to essential services such as potable water, drainage systems, adequate rubbish collection, basic sanitation and electricity contribute to the contamination of the domestic environment, which directly affects children's health. The absence of these resources, especially in rural areas and indigenous communities where infrastructure is limited, creates an environment prone to the proliferation of infectious diseases and a general deterioration in the quality of life.

  In these areas, the lack of clean water and adequate sanitation facilitates the spread of gastrointestinal diseases, which in turn significantly impacts children's growth and development, perpetuating the cycle of poverty and social exclusion. Lack of electricity and other essential services also limits opportunities for education and improved living conditions, affecting not only physical health, but also children's cognitive and emotional development.

- **Risk factors in the community**

Undernutrition, in addition to being related to physiological factors, is also influenced by a range of interacting social, political, economic, cultural and environmental factors. Poverty has long been recognised as a major underlying cause of undernutrition. According to (Chimborazo Bermeo & Aguaiza Pichazaca, 2023)poverty leads to low food availability and unequal access to food within the family, combined with problems such as overcrowding, lack of sanitation and inadequate child care. These conditions create an environment prone to the development of malnutrition, especially in young children, whose well-being depends to a large extent on the quality of the family and community environment.

According to (Naranjo Castillo A. E., 2020)these factors together contribute directly to the onset of protein-energy malnutrition. Lack of food education and low parental schooling, linked to poverty, generate incorrect feeding practices in the first years of a child's life. The late introduction of food, in unhygienic conditions, in insufficient quantities and with little nutritional variety, is often guided by erroneous beliefs or lack of knowledge about proper nutrition. As a result, children do not receive the necessary amount of energy, protein, vitamins and minerals for healthy growth, which depletes their nutritional reserves and increases their susceptibility to infections. This impacts not only their physical growth, but also their cognitive and emotional development, exacerbating the prevalence of protein-energy malnutrition.

The problem of undernutrition is not limited to the early years of life, but often originates from pregnancy. When the mother does not maintain adequate nutrition before or during pregnancy, or suffers from illness, she is more likely to give birth to an underweight baby, which increases the risk of developmental delay and undernutrition in the first months of life. According to (Alvarez Ortega, Desnutrición infantil, una mirada desde diversos factores, 2019)another factor that aggravates this situation is that many mothers, especially in contexts of poverty, are forced to devote much of their time to ensure the food security of the family, which reduces the time and attention they can give to their children, affecting their nutrition and care.

In addition to these individual and family factors, structural problems in the political and economic systems also play a key role. Paternalism and lack of community awareness limit the fair distribution of income, perpetuating inequality in access to essential resources. Inefficient health systems, which often lack the necessary equipment, resources and personnel, also contribute to undernutrition by preventing adequate and timely care for at-risk children. These systemic shortcomings create an additional barrier in the fight against undernutrition by limiting access to essential health services that could prevent and treat nutritional deficiencies in the most vulnerable communities.

- **Pathophysiology of child undernutrition**

Nutrition is closely related to the biological process of growth, which can be expressed by an increase (positive balance), maintenance (neutral balance) or decrease (negative balance) in body mass and volume, as outlined. These changes affect not only body size, but also adaptation to the demands of developing body form, function and composition. Adequate growth depends on a balance between synthesis and degradation of body tissues. However, when the rate of synthesis is less than the rate of destruction, a progressive loss of body mass occurs. This negative balance, which can have multiple causes, cannot be sustained for prolonged periods of time without causing serious, life-threatening organ dysfunction.

Malnutrition, as a state of chronic nutritional deficit, interferes with cellular functions gradually and systematically. Initially, the body's nutrient stores are compromised, affecting vital processes such as reproduction and growth. As malnutrition progresses, the body's ability to respond to stress, alter its energy metabolism and regulate intra- and intercellular communication mechanisms is impaired. These alterations also affect energy production and body heat generation, placing the body in an extreme catabolic state. Without timely intervention, this progressive catabolism inevitably leads to the degradation and eventual destruction of the individual.

The pathophysiology of malnutrition may involve four main mechanisms that contribute to the deterioration of the body:

a. Lack of energy intake: Insufficient intake of calories and essential nutrients prevents the body from receiving the energy needed for its basic functions, leading to reduced growth and tissue maintenance.

b. Disturbances in absorption: Even when food is ingested, if there are disturbances in the intestine's ability to absorb nutrients (such as in cases of chronic diarrhoea or intestinal disease), the body does not receive the necessary resources to support its functions.

c. Exaggerated catabolism: Catabolism is the process of breaking down complex molecules into simpler components to release energy. However, when this process is excessive, as in situations of stress or illness, the body destroys its own muscle mass and adipose tissue for energy, exacerbating the loss of body mass.

d. Excess excretion: When the body loses an excessive amount of nutrients through excretion, as in the case of kidney or gastrointestinal diseases, an imbalance is created that prevents the maintenance of an adequate nutritional status.

As these mechanisms fail, the body suffers a series of cascading effects that impact all systems, from immunity to the central nervous system, putting a child's life at serious risk. Child malnutrition not only affects physical growth, but also compromises cognitive and emotional development, perpetuating a cycle of poverty and vulnerability that is difficult to break without early and appropriate interventions.

- **Universal signs**

There are certain clinical signs that are present in all patients suffering from malnutrition, three main ones being observed in most cases:

a. Biochemical dilution: This phenomenon is especially common in cases of protein-energy malnutrition, where serum hypoproteinaemia is a characteristic feature, although it can also occur in other clinical pictures related to malnutrition. (Santafé Sánchez, Sánchez Rodríguez, A.L., & C.H., 2012).. Hypoproteinaemia manifests with a decrease in serum osmolarity, leading to electrolyte imbalances such as hyponatraemia (low blood sodium), hypokalaemia (low potassium) and hypomagnesaemia (low magnesium). These imbalances compromise the functioning of body systems, affecting vital processes such as nutrient transport and water regulation, which aggravates the patient's general condition.

b. Hypofunction: Under conditions of malnutrition, most of the body's systems show impaired functioning. This means that, due to a lack of essential nutrients, organs are unable to carry out their functions normally. This functional deficit affects key systems such as the cardiovascular, immune and digestive systems, reducing the body's ability to cope with infections, regulate body temperature and metabolise food. Hypofunction is also associated with reduced physical and mental activity, which impacts cognitive and motor development, especially in children.
c. Hypotrophy: The decrease in caloric intake causes a progressive depletion of the body's energy reserves. Initially, the body uses its glycogen and fat stores to compensate for the lack of nutrients, but over time, muscle mass is also affected, resulting in a significant loss of weight and height. This wasting not only manifests itself in physical appearance, such as a reduction in adipose panniculus (subcutaneous fat tissue) and muscle mass,

but also compromises the ossification process, slowing down bone growth (UNICEF, 2021). In children, this sign is of particular concern, as it affects their height and weight in relation to normal growth standards, which can have permanent consequences on their development if not treated in time.

The presence of these universal signs reflects the severity of the patient's nutritional status and demonstrates the multiple ways in which malnutrition impacts the body. Each of these signs highlights a critical aspect of the pathophysiology of malnutrition, showing how lack of nutrients affects not only physical growth, but also the body's functional and metabolic capacity. Without proper and timely intervention, these signs can progress to serious, life-threatening complications.

Psychomotor Development in Early Childhood

Psychomotor development during early childhood is an essential process that spans the first years of a child's life, a period in which the foundations for physical, cognitive, emotional and social growth are laid. This period, from birth to age six, is especially critical because the brain and nervous system are at their most plastic. This means that the stimuli and experiences received at this stage will have a profound and lasting impact on a child's future development. However, malnutrition can significantly disrupt this process, preventing proper development and causing delays that affect all areas of growth.

1. Importance of psychomotor development and its relationship to nutrition.

Psychomotor development is fundamental for children to acquire the skills necessary to interact effectively with their environment. During the first years of life, children progress through a series of stages in which they develop gross and fine motor skills, such as crawling, walking and manipulating objects. These skills are not only essential for physical autonomy, but are also deeply connected to cognitive and emotional development. However, when a child suffers from malnutrition, these stages can be compromised, leading to delays in the acquisition of motor, cognitive and social skills.

Adequate nutrition is essential for optimal brain function. During the first years of life, the infant brain develops rapidly, creating neural connections that are critical for learning. Lack of key nutrients such as protein, essential fatty acids and micronutrients (iron, zinc, iodine) can disrupt this process, resulting in poor psychomotor development. For example, a malnourished child who does not receive an adequate amount of food will not only have difficulty manipulating objects or performing motor tasks, but will also experience problems with attention, memory and cognitive processing, affecting their ability to learn and solve problems.

2. Components of psychomotor development and effects of malnutrition

Psychomotor development in early childhood encompasses several important components, all of which can be affected by undernutrition:

a. Motor development: Motor development involves the acquisition of gross and fine motor skills. Children suffering from malnutrition often have muscle weakness and lack of energy, which prevents them from progressing in key milestones such as crawling, walking or running. Lack of essential nutrients affects bone and muscle growth, which also delays finer skills such as grasping objects, drawing or performing tasks that require motor coordination.
b. Cognitive development: Cognitive development includes children's ability to think, learn and solve problems. In cases of malnutrition, the brain does not receive the nutrients it needs to develop neural connections efficiently. This can result in concentration difficulties, memory problems and a reduced ability to understand basic concepts such as space and time. These cognitive deficits can have a lasting impact on academic performance and intellectual development.
c. Emotional and social development: Malnourished children often experience difficulties in emotional regulation and social skills. Chronic malnutrition can lead to irritability, fatigue and a reduced ability to interact effectively with peers. In addition, the lack of an adequate nutritional environment during early childhood can limit their ability to form secure emotional bonds with their caregivers, affecting their long-term emotional and social well-being.
d. Language development: Malnutrition also has a direct impact on language development. Children who do not receive adequate nutrition during the first years of life may experience delays in language acquisition and the ability to form complex sentences or

express ideas. Lack of protein and key micronutrients affects brain development, which in turn impacts communication skills.

3. The impact of undernutrition on overall child well-being

Undernutrition in early childhood not only affects psychomotor development, but also has long-term repercussions on a child's physical and mental health. Undernourished children tend to have a weakened immune system, making them more susceptible to infectious diseases, and are less able to cope with emotional and social stress. At the cognitive level, a lack of essential nutrients during the early years can permanently affect academic performance, limiting opportunities for learning and success in school.

At the physical level, undernutrition compromises growth, which can lead to problems of short stature or stunted development. In addition, children who do not achieve adequate motor development due to undernutrition have greater difficulty performing everyday tasks such as writing or playing, which impacts their self-esteem and ability to integrate socially. Early nutritional support is therefore crucial to ensure that children build a solid foundation to enable them to develop holistically and face life's challenges.

Psychomotor development in early childhood is therefore an essential process that affects all aspects of children's well-being and growth. However, malnutrition severely compromises this development, affecting motor, cognitive, emotional and language areas. To ensure that children reach their full potential, it is critical that caregivers, educators and health authorities provide not only a stimulating and nurturing environment, but also adequate access to quality nutrition. Only in this way can optimal psychomotor development and a healthy future for children be ensured.

Effects of Malnutrition on Psychomotor Development

Undernutrition has a profound impact on children's psychomotor development, compromising the development of motor, cognitive, social and emotional skills. The human growth process requires adequate nutrition for children to reach psychomotor developmental milestones. Some of the most prominent effects are explained below:

a. Motor skills and coordination

- Reduced manipulative skills: Children suffering from malnutrition often exhibit muscle weakness and fatigue, which affects their ability to perform basic tasks involving manipulation of objects, such as picking up tools, drawing or playing with toys. This decreases hand-eye coordination and manual dexterity, key elements in psychomotor development. Lack of energy affects children's ability to control fine and gross movements, delaying their progress in fundamental activities such as walking, running or using utensils.
- Reduced response to stimuli: Malnutrition also compromises the central nervous system, resulting in slower motor response. This prevents children from reacting quickly to environmental stimuli, affecting both their ability to defend themselves and their

interaction with the physical environment. For example, a malnourished child may take longer to learn to run or jump, which impairs their ability to explore the world around them.

b. Cognitive growth and abstract thinking.

- Impaired abstract thinking: Cognitive development is also profoundly affected by malnutrition. For a child to develop problem-solving and planning skills, adequate nutrition is required to enable the creation and strengthening of neural connections. Malnourished children often have difficulty with tasks that involve abstract thinking, such as anticipating the outcome of an action or coordinating complex movements. This cognitive deficiency limits their ability to learn effectively and negatively affects their psychomotor development.

- Social learning delays: Social learning, which includes observing and imitating the actions of others, is key to psychomotor development. However, malnourished children, due to their lack of energy and delays in cognitive development, often have difficulty following and replicating the actions of peers and adults. This affects the acquisition of fine and gross motor skills, which are fundamental to a child's physical and social development.

c. Adaptation to the environment and survival

- Developmental impairment due to lack of nutrients: Malnutrition directly affects survival and the development of skills needed for daily life. Children who do not receive adequate nutrition have less energy to learn to move, manipulate objects and explore the environment. This compromises their ability to adapt to different situations, affecting their ability to feed themselves adequately and to develop other psychomotor skills necessary for survival.

- Changes in activity patterns: Malnourished children tend to be less active due to chronic fatigue and lack of energy. This impacts their ability to maintain regular physical activity patterns, which may limit their ability to develop advanced motor skills. In the long term, this can alter their psychomotor development, affecting their ability to participate in both physical and mental activities.

d. Social and community development

- Difficulties in interaction and communication: Malnutrition can affect language development and social skills, limiting children's ability to interact with peers and caregivers. Lack of adequate nutrition delays the acquisition of language and body language skills, essential components of effective communication. Malnourished children tend to show less interest in social interaction, which prevents them from learning through observation and play, critical activities for their psychomotor and social development.

- Inhibition of play and socialisation: Play, an essential tool for psychomotor development, is restricted in malnourished children due to lack of energy. By not actively engaging in play, children miss opportunities to develop crucial social and motor skills. Activities involving body expression, use of gestures and group coordination are limited, making it difficult for them to adapt socially and participate in the community environment.

e. Culture and traditions

- Limitations in participation in cultural and symbolic activities: In many cultures, symbolic and ritual activities, such as traditional dances or ceremonies, involve coordinated and rhythmic movements that promote psychomotor development. However, malnourished children are often too weak to participate in these activities, which affects their cultural integration and the acquisition of psychomotor skills related to rhythmic expression and group coordination.

- Impact on sense of belonging and psychomotor development in cultural contexts: The inability to participate fully in rituals and traditions affects not only the psychomotor development of malnourished children, but also their sense of cultural identity and belonging. The practice of symbolic movement and coordination in the context of traditional activities promotes both physical and emotional development, which is limited in chronically malnourished children.

Malnutrition has a devastating impact on children's psychomotor development. By affecting motor as well as cognitive and social skills, it limits their ability to interact with their environment, learn effectively and participate in activities that are key to their overall well-being. Combating malnutrition not only improves children's physical health, but also enables them to reach their full potential in terms of psychomotor, cognitive and social development, thus giving them the opportunity to better integrate into their community and culture.

Nutritional Interventions

According to the World Health Organisation (WHO), malnutrition affects more than 200,000 children under the age of five worldwide, who do not have enough food to meet their basic needs. This phenomenon is alarming, as the lack of essential nutrients at such a critical stage of development has serious and long-term consequences on children's physical and cognitive health. Among the many alternatives to combat malnutrition, one of the most promising is the use of soy as a primary source of protein. (Health O. M., 2021)..

Soy is a highly efficient legume for the production of good quality protein, and is a viable option in a world where food shortages are a growing problem. Combining soy protein with protein from cereals offers an effective solution to increase both the quantity and quality of protein in the diet. This is because the proteins from both food groups are complementary. Cereals are often deficient in lysine, an essential amino acid that is present in soy. This nutritional synergy makes it possible to improve the biological value of the proteins consumed, which is key to combating malnutrition. In addition, soy can be grown in large quantities, making it a sustainable and accessible resource for mitigating malnutrition in various regions of the world. (Alberto Javier García-Garroa, 2007)..

Severe Acute Malnutrition (SAM) and its relation to infections

Malnutrition, in particular Severe Acute Malnutrition (SAM), is generally associated with an increased incidence of various infections, such as gastroenteritis, acute respiratory infections, HIV/AIDS and tuberculosis, among others. These infections not only aggravate the condition of affected children, but also complicate the clinical management of malnutrition, as patients may require complex treatments including second-line antibiotics, antiretrovirals or TB-specific drugs. However, the pharmacokinetics of these drugs in malnourished children has been insufficiently studied. In patients with DAS, the physiological system is compromised, which alters the disposition, metabolism and efficacy of drugs, which in turn affects the treatment of infections (Goyheneix, 2019).

Improved research on the management of these drugs in undernourished populations is crucial, as drug administration may present additional risks in children with weakened immune and metabolic systems. Correct dosing and selection of drug treatments is essential to avoid complications and ensure treatment effectiveness.

Nutritional treatments for malnutrition

A comprehensive approach to the treatment of malnutrition should focus on ensuring that patients adequately tolerate the food and nutrients they receive, as well as ensuring their bioavailability. Effective nutritional management should preserve the balance of the gut microbiome and reduce levels of systemic inflammation, as these factors are key to overall health recovery.

Foods for special medical purposes play an important role in the treatment of malnutrition. These foods are specifically formulated to meet the dietary needs of patients who are unable to ingest, digest, absorb or metabolise normal foods due to their medical condition. They are essential for patients with limited ability to process conventional foods, including infants who require medical supervision for dietary management. (Moreno, Señarís, & Montserrat, 2020)..

These specialised products not only improve nutritional status, but also help restore normal bodily functions by providing nutrients in forms that are easier to absorb. This is especially important in severely malnourished children, who may have digestive or metabolic problems that make recovery difficult if they are given foods not adapted to their specific needs.

Increasing sustainability and accessibility of nutrition intervention

In addition to traditional approaches to nutrition intervention, it is critical to consider the sustainability and accessibility of proposed solutions. The mass production of crops such as soy, coupled with initiatives to improve the distribution of specialised foods in vulnerable areas, offers an opportunity to address the problem of undernutrition in a more efficient and long-term manner. Implementing agricultural policies that promote the cultivation of protein-rich foods such as soy can make a significant difference in preventing child undernutrition in rural and impoverished areas.

At the same time, there is a need to improve the health infrastructure in these regions to ensure that malnourished children receive not only adequate nutritional treatment, but also the medical care necessary to manage infections associated with malnutrition. With better coordination between agricultural policies, health services and feeding programmes, the impact of child malnutrition globally could be significantly reduced.

Feeding and nutritional recovery programmes

Feeding and nutrition recovery programmes are key interventions designed to combat malnutrition, especially in vulnerable populations such as children under five, pregnant women, infants and low-income communities. These programmes aim not only to provide access to quality food, but also to restore the nutritional status of people already suffering from malnutrition, promoting their comprehensive and sustained recovery.

a. Objectives of the Food Programmes

The main objective of feeding programmes is to ensure that individuals receive the essential nutrients necessary for their development and well-being. In the case of children, these programmes seek to prevent and treat acute and chronic malnutrition, which can have devastating effects on their physical growth, cognitive development and overall health. Feeding programmes are also designed to ensure food security by ensuring that vulnerable families have consistent access to nutritious food, even in times of crisis such as natural disasters or conflict situations.

b. Key Components of Nutritional Recovery Programmes

1. Distribution of Fortified Foods: Fortified foods, such as cereals fortified with vitamins and minerals, are frequently used in these programmes to ensure that people receive essential micronutrients that are often lacking in their diets. This is especially important in rural and impoverished areas, where access to a varied diet is limited.

Therapeutic Nutritional Supplements: In cases of severe acute malnutrition, therapeutic nutritional supplements are essential for recovery. These products, such as fortified peanut pastes and special formulations, are designed to provide a high concentration of calories and easily absorbed nutrients,·allowing for rapid recovery of the nutritional status of the most critically ill patients.

3. Comprehensive Medical Care: Nutritional recovery programmes often include comprehensive medical care to treat complications associated with malnutrition, such as gastrointestinal infections, immune deficiencies, anaemia and respiratory diseases. This holistic approach is essential, as malnutrition weakens the immune system, making patients more susceptible to infection and disease.

Nutrition Education: Nutrition education is a key component to ensure that interventions are sustainable in the long term. Teaching families and communities about the importance of a balanced diet, hygiene practices and proper child feeding methods is crucial to prevent the recurrence of malnutrition. This component also fosters the empowerment of communities to better manage their food resources.

Types of Feeding Programmes

There are different types of feeding programmes, tailored to the specific needs of the populations they serve. Some of the most common are:

1. School feeding programmes: These programmes provide nutritious meals to school-aged children, improving not only their nutritional status but also their academic performance. School meals can be the main source of quality food for many children living in poverty.
2. Nutritional recovery centres: These centres are designed to treat severe cases of malnutrition, providing not only food, but also medical care and constant monitoring to ensure the proper recovery of affected children and adults.
3. Emergency food assistance programmes: During crises such as wars, natural disasters or pandemics, food assistance programmes provide basic food to affected communities, thus preventing mass starvation and malnutrition.

Importance of Sustainability

One of the greatest challenges facing feeding and nutrition recovery programmes is ensuring their long-term sustainability. For these interventions to be effective, they must be accompanied by policies that promote continued access to quality food and encourage local food production. Collaboration between governments, non-governmental organisations and the community is essential to create efficient and equitable food distribution systems.

In addition, it is crucial to invest in health infrastructure and education so that people can learn to manage their own food security. Programmes should work on local capacity building, so that communities can maintain good nutrition without relying on external aid in the future.

Therefore, feeding and nutrition recovery programmes play a vital role in the fight against malnutrition worldwide. By providing access to nutritious food, therapeutic supplements and medical care, these programmes help restore the health of the most vulnerable people and prevent the long-term consequences of undernutrition. To maximise their impact, these programmes need to be sustainable and integrated into a broader strategy that includes nutrition education and community strengthening. Only through a comprehensive approach can we ensure that all children and adults have the opportunity to develop to their full potential and live healthy lives.

Popular education's perspective on nutrition education

Through the implementation of health education, the misinterpretation of malnutrition could be eradicated, as it is possible to develop a proper relationship with food and explain adequate intakes to prevent potential health problems. However, new pedagogical practices such as nutrition education are necessary for the integral recognition of the person, and thus understand their previous experiences, in order to facilitate the incorporation of useful tools and healthy foods in their daily lives. (Ruiz Arciniega, 2021)..

When engaging in nutrition education, information about proper diets, methodologies for good nutrition and healthy diets, as well as healthy food preparation should be taken into account. Through the information provided on nutrition, health and social rights, it is estimated that people create awareness about the problem of malnutrition that is being created and in turn expanding worldwide, it is expected that citizens promote actions of citizen participation in the construction of public policies on nutrition and health, in order to implement healthy habits and lifestyles and thus improve the quality of life of each of them. (Quevedo Bolívar, 2019)..

Prevention and reduction of child malnutrition

Therefore, the prevention of child undernutrition requires both short- and long-term approaches, as well as improving the social and economic determinants of undernutrition, which include improving maternal education, providing economic opportunities

to grow and purchase food for children, water and sanitation, access to quality health services and empowerment of women in society.
While, in the short term, reducing child undernutrition requires the protection, promotion and support of breastfeeding, providing counselling and education on complementary feeding especially in the absence of food security and the provision of complementary foods along with appropriate counselling, reducing the frequency and duration of infections and diarrhoea, and the promotion of increased food intake after illness for the "rapid growth" phase.

To achieve adequate coverage, the interventions described above must be integrated into primary health care programmes for pregnant women, childbirth, newborn and young child care. Improving infant and young child nutrition should be a priority for all health workers and should not be limited to the domain of nutritionists.

Health workers must have the knowledge and technical skills to provide appropriate counselling to mothers on breastfeeding and complementary feeding; manage feeding and nutrition problems; and treat diseases that lead to malnutrition. (Saca Molina, 2018).

The following aspects should be considered:
- **Improvement of diet:** promotion of breastfeeding, adequate complementary feeding and feeding during illness and convalescence.

- **Micronutrient supplementation:** Vitamin A supplementation, iron supplementation and zinc supplementation.

- **Public health actions:** growth promotion and monitoring, provision of basic health services and safe water use (Ministry of Public Health, 2009).

According to (Carpio-Arias, et al., 2018) beyond its often painful bodily expression and its affective and emotional content, nutritional diseases have a social meaning and significance, as a pathological expression of the needs of those who suffer from them, affecting everyone. Malnutrition is therefore considered as a phenomenon of multifactorial origin, resulting from a wide range of social and economic conditions, and is one of the most pressing problems in the child population.

- Innovative projects in different communities

One of the main components that exists in the population that has more weight in the prevention of malnutrition is the educational component, where we as professionals propose this article for parents to take into consideration the formal education and nutritional health education where different practices that are specifically designed to protect and promote health will be detailed, an example is to provide women with abundant food, dense in energy, during the first months after childbirth, in order to promote and protect the health status of the community.

Malnutrition is more present in times of maximum acceleration of children's growth, especially in infants from 0 to 5 years old.

Therefore, as professionals we need to know the nutritional status of these children in Guayaquil, depending on their age, sex, eating habits and associated components such as socioeconomic status, so that through these data we can plan prevention activities and procedures for the benefit of the population, thus running the risk of severely influencing the central nervous system, the protective system of the body (immune) and the psychomotor and psychosocial development of children under 5 years of age. (García-Sámano, 2018).
In addition, it is necessary to contribute to Primary Health Care by providing a statistical basis on the issue of malnutrition, which is currently considered a public health problem for the prevention of malnutrition. One way to overcome this problem is by applying a nutritional rehabilitation programme that not only reverses this physical problem, but also promotes behavioural changes in the nutrition and health of the mothers of infants with nutritional deficits, with the aim of obtaining different benefits that will not only allow the infant to recover nutritionally, but will also prevent this or any other infant in society from falling into the clutches of malnutrition.

On the other hand, it is necessary to place more emphasis on this problem, since in our country there are many communities with low economic resources, where we will be able to help them and their families, since this leads to long-term problems at a social level, where the only way to counteract child malnutrition is to attack directly the cause of this social problem due to the size of the nutritional problem in Ecuador, It is essential to have an effective system to detect children with malnutrition and to rehabilitate them in a correct, efficient and comprehensive way that makes it possible to sustain them in time and that involves not only health personnel but also the mothers of malnourished children and society as a whole.

Incidence of child malnutrition in under-fives in Ecuador

According to the World Food Programme (WFP), Ecuador is considered the fourth country in Latin America, after Guatemala, Honduras and Bolivia, with the worst rates of child malnutrition. Currently, 26% of the Ecuadorian child population, from birth to five years of age, suffers from chronic malnutrition, a situation that is worse in rural areas, where it reaches 35.7% of children, and is even more critical in Amerindian children, with rates of more than 40%. In Latin America, almost 40% of families live in extreme critical poverty, approximately 60 million children belong to these families; 20% of these families live in a state of absolute poverty. (Peñafiel Carriel, 2015)..

Undernutrition is mainly due to insufficient protein and micronutrient intake, there are approximately 821 million undernourished people in the world, which is the main cause of stunting, low height and cognitive deficits and in developing countries is associated with 60% of child deaths (Rivadeneira, 2020).. However, micronutrient deficiencies, iron deficiency is more prevalent.

In Mexico, 4.4 per cent of the population has reported extreme poverty and 20 per cent below minimum welfare. In children with moderate acute malnutrition, supplementary foods produce an average increase in the rate of recovery of close to 30 % (0.28); moreover, when these foods are withdrawn, relapse rates reach 50 % in the first year. (Talavera, 2020)..

Summary of key findings

In the first place, it could be said that in most cases, mothers are not well informed about adequate food for their children and in turn are influenced by the environment that surrounds them, such as their parents' low income, their parents' low academic level, and other factors such as housing, basic services and, above all, having good access to food: Low economic income of their parents, low academic level of the same, among other factors that influence as to have housing, basic services and above all to have a good access to food.

The aim of this project is to train people in this sector of the city in the importance of good nutrition in the infancy stage, since, as has been mentioned, this is a stage in which children need to be fed very well, as it is at this stage that their bodies need more than others to obtain the nutrients and vitamins that are necessary for excellent development.

Based on the fact that we have the necessary and sufficient information that allows us to reach the following conclusion, which states that if there is a remarkable inter-institutional coordination, it will be possible to carry out awareness campaigns regarding chronic child malnutrition and to contribute to its improvement.

Personalised training and ongoing support for breastfeeding mothers increases the percentages of children fed exclusively during the first 3 months of life, which helps us to reduce the percentage of malnutrition in the country.

Proposals for future studies and interventions

Early and automated nutritional intervention coupled with clinical collaboration is critical to remedy the issue of malnutrition in hospitals and has great potential to improve patient care as well as reduce hospital costs.

Breast-feeding should always be continued and supplemented as indicated in the dietary guidelines for children under two years of age. Nutritional treatment such as formulas should be given gradually supplemented with other foods, especially those that may be available in the patient's home.

To implement administrative and care policies that favour an institutional culture that ensures the prevention of malnutrition and undernutrition in patients.

To put in place processes to ensure adequate nutritional intervention and that the nutritional status of patients is routinely monitored.

References

Alberto Javier García-Garroa, M. G.-F.-O. (2007). Soy treatment of malnourished patients aged 1 to 4 years. *Elseiver*, 69-73 .

Alvarez Ortega, L. G. (2019). Child malnutrition, a look from different factors. *Investigación Valdizana, 13(1),* , 15-26.

Alvarez Ortega, L. G. (2019). Child malnutrition, a look from different factors. *INVESTIGACIÓN VALDIZANA*, 1-12.

Álvarez Ortega, L. G. (2019). Child malnutrition, a look from different factors. *Universidad Nacional Hermilio Valdizán, Peru,* 15-26.

Álvarez Ortega, L. G. (2019). Child malnutrition, a look from different factors. *Universidad Nacional Hermilio Valdizán*, 15-26.

Brahm, P., & Valdés, V. (2017). Benefits of breastfeeding and risks of not breastfeeding. *Rev Chil Pediatr*, 7-14.

Cancela, M. d. (15 November 2021). *Innatia - Health, well-being and traditions* . Retrieved from http://www.innatia.com/s/c-alimentacion-infantil/a-grados-de-desnutricion-en-los-ninos.html

Carpio-Arias, T., Ramos-Padilla, P., Delgado-López, V., Villavicencio-Barriga, V., Carpio-Salas, J., & Morejón-Terán, Y. (2018). NUTRITIONAL STATUS, FOOD CONSUMPTION, PHYSICAL ACTIVITY AND EATING DISORDERS IN ADOLESCENTS FROM URBAN AND RURAL AREAS IN THE ANDEAN REGION OF ECUADOR. *Talentos revista de invetigación*, 84-93.

Chimborazo Bermeo, M. A., & Aguaiza Pichazaca, E. (2023). *Factors associated with chronic child undernutrition in children under 5 years of age in Ecuador: A systematic review.* Cuenca: LATAM.

Connect, E. (18 December 2018). Nutritional diseases (structural and functional pathology): marasmus and kwashiorkor. *ELSEVIER*, pp. 1-5.

Cueva Moncayo, M., & Pérez Padilla, P. (2021). Child malnutrition in Ecuador: A literature reviewChild malnutrition in Ecuador. A review of the literature. *Boletín de Malariología y Salud Ambiental*, 556-564.

Cuevas-Nasu, L., Gaona-Pineda, E. B., Rodríguez-Ramírez, S., Morales-Ruán, M., González-Castell, L., García-Feregrino, R., . . . Rivera-Dommarco, J. (2019). Desnutrición crónica en población infantil de localidades con menos de 100 000 habitantes en México. *Salud pública de mexico*, 6-9.

Fernández-Martínez, L. S.-L.-L.-C.-D.-M. (02 of 2022). Determining factors in child malnutrition in San Juan and Martinez, 2020. *Scielo Rev Medical Sciences*, pp. http://scielo.sld.cu/scielo.php?script=sci_arttext&pid=S1561-31942022000100005.

García-Sámano, V. M. (2018). Chronic Childhood Undernutrition. . *ENSANUT*, 45 - 56.

González, Z. F., Font, A. J., Ochoa, M. Y., Rodríguez, E. C., & Estrada, A. M. (2020). Malnutrition; a global public health problem. *SCielo*, 1-5.

Jiménez Benítez D., R. M. (2010). Análisis de determinantes sociales de la desnutrición en Latinoamérica. https://www.researchgate.net/publication/268376586_Analisis_de_determinantes_sociales_de_la_desnutricion_en_Latinoamerica.

Martínez, J. G., Salazar Duque, D., Portugal Morejón, C., & Lala Gualotuña, K. (2020). Nutritional status of children under five years of age in the parish of Pifo. *Nutrición clínic y dietetica hospitalaria*, 90-99.

Martínez, R., & Andrés, F. (2007). *El costo del hambre: impacto social y económico de la desnutrición infantil en Centroamérica y República Dominicana.* Dominican Republic: Economic Commission for Latin America and the Caribbean (ECLAC).

Ministry of Public Health, and. A. (2009). MANUAL OF PROCESSES. *MSP*.

Moreno, S. P., Señarís, J. d., & Montserrat, J. G. (2020). Treatment of disease-related malnutrition: status of regulation in the Spanish National Health. *Nutrición Hospitalaria*, 1246-1280.

Naranjo Castillo, A. E. (2020). *Kwashiorkor child malnutrition.* Guayaquil: RECIMUNDO.

Naranjo Castillo, A. E., Alcivar Cruz, V. A., Rodriguez Villamar, T. S., & Betancourt Bohórquez, F. A. (03 2020). Child malnutrition kwashiorkor. *Saberes del Conocimiento*, pp. 24-45.

Paraje, G. (2008). Evolution of chronic child undernutrition and its socio-economic distribution in seven countries in Latin America and the Caribbean. *ECLAC/UNICEF*.

Peñafiel Carriel, A. N. (2015). *"INCIDENCE OF MALNUTRITION AND ITS RELATIONSHIP WITH ANAEMIA IN CHILDREN UNDER 5 YEARS OF AGE ATTENDING THE CLINIC OF THE SOCIAL SUPPORT FOUNDATION FATHER MANUEL SESMA, PALENQUE.* Quevedo: UNIVERSIDAD TÉCNICA ESTATAL DE QUEVEDO.

Pozo Aguilar, M. S. (10 November 2022). SOCIODEMOGRAPHIC CONDITIONS AND NUTRITIONAL STATUS IN CHILDREN AGED 1 TO 5 YEARS IN THE PARISH OF ANGOCHAGUA. *Repositorio*. Ibarra, Ecuador: UNIVERSIDAD TÉCNICA DEL NORTE.

Quevedo Bolívar, P. A. (2019). Malnutrition: beyond nutritional deficiencies. *Department of Social Work, Faculty of Human Sciences, National University*, 219-239.

Quirindumbay Uchupailla, S. M. (12 2016). *Child malnutrition and the problems it generates in the neuromotor development of children aged 1 to 3 years in the CNH San Fernando, province of Azuay, canton San Fernando*. Retrieved from Repositorio Institucional de la Universidad Politécnica Salesiana: https://dspace.ups.edu.ec/handle/123456789/13173

Ramos-Padilla, P. D.-L.-B.-A. (06 April 2020). Nutritional typologies in children under 5 years of age in the province of Chimborazo, Ecuador. *Spanish Journal of Human Nutrition and Dietetics*, pp. 1-5.

Realpe Muñoz, A. M. (2013). SEVERE KWASHIORKOR-TYPE MALNUTRITION. *Revista Gastrohnup*, pp. 20-26.

Rivadeneira, M. F. (2020). A multicausal model of chronic malnutrition and anaemia in a rural coastal child population in Ecuador. *Revista de salud maternoinfantil* , 472 - 482.

Ruiz Arciniega, J. G. (2021). Child malnutrition and its effect on neurodevelopment: a critical review from an Ecuadorian perspective. *Mikarimin. Multidisciplinary Scientific Journal*, 131-146.

Saca Molina, E. C. (September 2018). *Repositorio Universidad Técnica de Cotopaxi* . Retrieved from https://repositorio.utc.edu.ec/items/3b2670fe-a62d-47dc-a5fb-1d6f6055464c

Health, O. M. (2021). Levels and trends in child undernutrition: Joint UNICEF/WHO/World Bank Group estimates of child undernutrition: key findings from the 2021 edition. *https://iris.*who.int/handle/10665/341135, p. 31.

Health, O. M. (2024). Malnutrition. *OMG*, 1-6.

Santafé Sánchez, L. R., Sánchez Rodríguez, D., A.L., V. G., & C.H., G.-C. (2012). Nutritional status among hospitalized children with mixed diagnoses. *Nutrición Hospitalaria* , 1451-1459.

Senna Rodrigues, B., Campos Pellanda, L., & Andreatta Gottschall, B. (2012). NUTRITIONAL ASSESSMENT OF CHILDREN AND TEENAGERS. *Rev Chil Nutr Vol. 39, Nº2*.

Talavera, J. O.-V.-A.-A.-F.-J.-E. (2020). Prevention of moderate acute malnutrition with a ready-to-eat food supplement in preschool children in rural communities. *Scielo*, 509-518.

UNICEF. (2021). Levels and trends in child malnutrition. *UNICEF, WHO and the World Bank Group*, pp. 4-81.

UNICEF, p. l. (2022). Chronic Child Undernutrition. 1-6. Retrieved from https://www.unicef.org/ecuador/sites/unicef.org.ecuador/files/2021-03/Desnutricion-Cronica-Infantil.pdf

MIX
Papier aus verantwortungsvollen Quellen
Paper from responsible sources
FSC® C105338

Printed by Books on Demand GmbH, Norderstedt / Germany